Naming Compounds Chemistry

A High School Chemistry Workbook with Answers
(250+ Compounds to Practice)

Simple Ionic Compounds

Naming Simple Ionic Compounds

1. Name the **metal** first.
2. Next, use the first syllable of the non-metal + "ide"

Example: **MgCl₂**   magnesium chloride

Determining Chemical Formulae of Ionic Compounds

To determine the chemical formula, we must use the **Zero Sum Rule**.

This means that the <u>sum of the positive charges must equal the sum of the negative charges</u>, resulting in a net charge of **zero** for the compound. This ensures that the compound will be electrically neutral.

	CaCl
Example: **Calcium chloride**	$\overset{+2 \; -1}{CaCl}$
1. Write the symbol for each element in the compound (metal first, then non-metal).	
2. Find the charge of the ion that each element will form.	$\overset{+2 \; -1 \, (x2)}{CaCl}$
3. Use the Zero Sum Rule to determine how many ions of each element are needed to give a net charge of zero. *Do not include subscripts of 1.	$\overset{+2 \; -2 \, = \, 0}{\underset{\overline{}}{Cl}}$

∴ 1 calcium ion will bond with 2 chloride ions, resulting in a chemical formula of **CaCl₂**. $\overset{+2 \quad -1}{Ca_x Cl_2}$ ∴ **CaCl₂**

💡 **Confused? Here's a helpful trick!**

The Criss-Cross Rule:
1. Write the symbol for each element with its ionic charge above it.
2. Take the charge from above the atom and use it as the subscript for the other atom in the compound (don't write the + or – signs; don't write a subscript of 1).
3. Divide the subscripts by a common factor if there is one (i.e. reduce!).

Practice: Chemical Formulae of Simple Ionic Compounds

Determine the chemical formula for each of the following compounds:

1) Magnesium sulfide
2) Sodium phosphide
3) Lithium oxide
4) Aluminum nitride
5) Potassium bromide
6) Strontium chloride

Simple Ionic Compounds

7) Sr₃P₂
8) BaI₂
9) NaBr
10) Rb₂O
11) Li₂S
12) MgBr₂

Table of Contents	Page

Instructions for Use

Learning to name inorganic chemical compounds is a *crucial skill* for any chemistry student. It is also a skill that becomes increasingly more complicated with every new rule that is presented. What seemed simple at first can quickly frustrate and overwhelm students.

This content is designed to scaffold the naming process in a way that has students learning one new rule at a time and then practicing it. From there, they will do *Mixed Practice* work pages that combine everything that have learned up until that point.

The workbook should be completed in the order it is presented, as each skill builds on what was learned before. The topics included in this workbook are:

- **Navigating the Periodic Table of Elements**
- **Simple Ionic Compounds**
- **Ionic Compounds containing Multivalent Metals**
- **Ionic Compounds containing Polyatomic Ions**
- **Molecular Compounds**
- **Binary Acids and Oxyacids**

For each topic, this workbook includes:

Short Lessons
Each new topic begins with a short lesson. It includes an explanation of the new chemical compounds and step-by-step instructions on how to name them and determine their chemical formulae.

Practice: Naming
Each lesson is followed by 12 straightforward questions where students must name a chemical compound given the chemical formula. Answers are found at the end of the workbook.

Practice: Determining Chemical Formulae
After naming practice, students determine the chemical formula of 12 different compounds when given their names. Answers are found at the end of the workbook.

Self-Checking Practice
Each section contains a self-checking practice worksheet, where students can practice their new skills to reveal a secret message or the answer to a riddle. They will know if they are right if these messages make sense! These self-checking practice worksheets are designed for students to be able to find and correct their own mistakes.

Mixed Practice
Each section is followed by *Mixed Practice*, which mixes each new skill into questions requiring naming compounds and determining chemical formulae. These worksheets combine everything that has been learned up until that point.

Periodic Table of Elements

The **periodic table** is a chart that organizes all the different **elements** that make up the universe. It helps us see how elements are connected and allows us to **predict and explain how they will act** in different situations.

The elements are organized by **atomic number**, which is the same as the number of **protons** and **electrons** in a neutral atom. The **atomic mass** is made up of the mass of protons + the mass of neutrons in an atom.

The periodic table is a tool, so _you don't need to memorize it_. You just need to know **how to use it**!

But as you work through this workbook (and the rest of your studies in chemistry) you'll probably start to see that you have memorized many of the elements and their symbols, locations, ionic charges, and more without even trying!

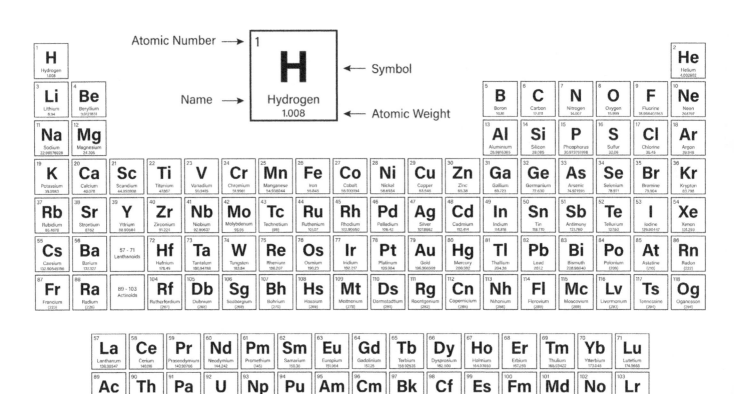

Navigating the Periodic Table of Elements

The periodic table is organized into **7 rows** called **periods** and **18 columns** called **groups**. Each period represents a new energy level for electrons, while groups contain elements with similar properties.

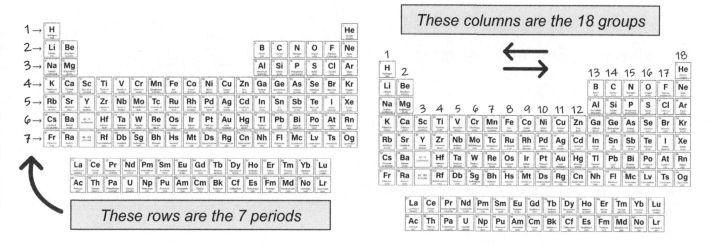

These rows are the 7 periods

These columns are the 18 groups

Some of the groups have special names that you should know. These are:

Group 1: Alkali Metals (very reactive, often bond with halogens)
Group 2: Alkaline Earth Metals (less reactive)
Group 17: Halogens (very reactive, often bond with alkali metals)
Group 18: Noble Gases (stable and unreactive)

It is also important to be able to identify if an element is a **metal** (found on the left side), a **non-metal** (found on the right side), or a **metalloid** (shown in gray; they border the dividing line between metals and non-metals). As you can see, the majority of elements are metals.

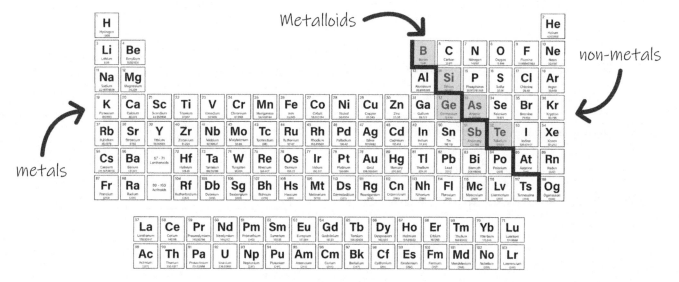

Metalloids

non-metals

metals

Practice: Who am I?

Use the clues below to identify each mystery element.

1) I have 12 protons. _____

2) I have 1 proton and 1 electron. _____

3) I am the Noble gas of the 1st period. _____

4) I am the Halogen of the 3rd period. _____

5) I am in group 4 and period 6. _____

6) I have 17 electrons. _____

7) I have 92 protons. _____

8) I am the Noble gas of the 4th period. _____

9) I am the lightest Alkaline Earth metal. _____

10) I am the Alkali metal in period 3. _____

11) I am a metalloid in group 13. _____

12) I am a metal with atomic mass of 40 amu. _____

Practice: Chemical Symbols

Match each element with its chemical symbol.

1) Sodium   Sn

2) Tin   K

3) Copper   Au

4) Mercury   Pb

5) Potassium   Cu

6) Lead   P

7) Antimony   Sr

8) Silver   Na

9) Gold   Sb

10) Strontium   N

11) Phosphorus   Ag

12) Nitrogen   Hg

Practice: Metal or Nonmetal?

*Determining if an element is a **metal** or a **nonmetal** is very important when it comes to naming chemical compounds.*

Take a look at the element names and symbols below.

Circle each metal. Underline each nonmetal and metalloid.

Cl   F   Be

lithium

                                                              magnesium

phosphorus   Cu

                                              H   S

Rb

        oxygen

                                        nitrogen

Ti

              tungsten   carbon

oxygen   Pb

            sulfur

                                                  helium

Si

Simple Ionic Compounds

Atoms and Ions

When atoms gain or lose valence electrons they become electrically charged and form ions.

The process of atoms becoming ions occurs when elements bond with other elements to form compounds.

Take a look at the example below:

$$Na + Cl \longrightarrow Na^+ + Cl^- \longrightarrow NaCl$$

Ionic Compounds (NaCl)

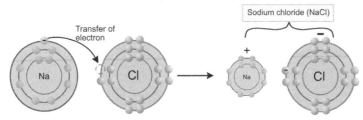

So, atoms are looking for opportunities to bond with other elements by transferring valence electrons from one element to another in such a way that they'll all have full outer shells and be stable.

When this happens, the **ions in the compounds will be attracted to each other.**

Most elements will always have the **same** predictable charge when they're part of an ionic compound (*some transition metals are an exception to this rule, but we'll get to that later…*).

Ionic charges are crucial in naming ionic compounds and determining their chemical formulae.

Recall: When in ionic compounds, elements of the following groups will always have these charges

Group 1	Group 2	Group 13	Group 14	Group 15	Group 16	Group 17	Group 18
+1	+2	+3	+/-4	-3	-2	-1	0

Simple Ionic Compounds

Naming Simple Ionic Compounds

1. Name the **metal** first.
2. Next, use the first syllable of the non-metal + **"ide"**

Example: **MgCl$_2$**

magnesium chloride

Determining Chemical Formulae of Ionic Compounds

To determine the chemical formula, we must use the **Zero Sum Rule**.

This means that the <u>sum of the positive charges must equal the sum of the negative charges</u>, resulting in a net charge of **zero** for the compound. This ensures that the compound will be electrically neutral.

Example: **Calcium chloride**

1. Write the symbol for each element in the compound (metal first, then non-metal).	CaCl
2. Find the charge of the ion that each element will form.	+2 -1 CaCl
3. Use the Zero Sum Rule to determine how many ions of each element are needed to give a net charge of zero. *Do not include subscripts of 1.*	+2 -1 (x2) CaCl Cl ———— +2 -2 = 0
∴ 1 calcium ion will bond with 2 chloride ions, resulting in a chemical formula of **CaCl$_2$**.	

 Confused? Here's a helpful trick!

The Criss-Cross Rule:
1. Write the symbol for each element with its ionic charge above it.
2. Take the charge from above the atom and use it as the subscript for the ***other*** atom in the compound (don't write the + or – signs; don't write a subscript of 1).
3. Divide the subscripts by a common factor if there is one (i.e. reduce!).

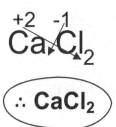

Practice: Naming Simple Ionic Compounds

Name each of the following compounds:

1) NaCl _____

2) CaF_2 _____

3) LiBr _____

4) Al_2S_3 _____

5) Mg_3N_2 _____

6) K_2O _____

7) Sr_3P_2 _____

8) BaI_2 _____

9) NaBr _____

10) Rb_2O _____

11) Li_2S _____

12) $MgBr_2$ _____

Practice: Chemical Formulae of Simple Ionic Compounds

Determine the chemical formula for each of the following compounds:

1) Magnesium sulfide   _____

2) Sodium phosphide   _____

3) Lithium oxide   _____

4) Aluminum nitride   _____

5) Potassium bromide   _____

6) Strontium chloride   _____

7) Rubidium phosphide   _____

8) Sodium iodide   _____

9) Beryllium oxide   _____

10) Aluminum fluoride   _____

11) Magnesium oxide   _____

12) Calcium phosphide   _____

Self-Checking Practice: Simple Ionic Compounds

Determine the chemical formula of each compound below. Find your answer on the next page and cross out the letter above it. When you finish, the answer to the question below will remain.

magnesium sulfide	calcium bromide
potassium chloride	sodium oxide
barium bromide	cesium fluoride
rubidium iodide	magnesium nitride
aluminum sulfide	calcium oxide
sodium sulfide	strontium bromide
potassium phosphide	

Self-Checking Practice: Simple Ionic Compounds

What kind of tree fits in your hand?

D	A	S	N	P	A	S	L	A	D	V
CaO	CsF_2	Mg_3N_2	MgS	Sr_2Br	NaS_2	CsF	Al_3S_2	Al_2S_3	$CaBr_2$	$SrBr_2$
L	Y	M	T	I	W	R	E	R	C	E
RbI	$BaBr_2$	Cs_2F_3	Rb_2I	KCl	K_3P	CaO_2	$BaBr$	Na_2S	Na_2O	KP_3

Answer: _____

Ionic Compounds (Multivalent Metals)

So far we've looked at how ionic compounds are formed from either **alkali metals, alkaline earth metals, or Group 13 metals + nonmetal**.

The rules are slightly different when we look at compounds that are formed from **multivalent metals + nonmetals**.

These metals are those found in the **centre** block of the periodic table and include elements such as **iron, chromium, nickel, tin** and **lead**.

There is no pattern to figure out what ions these metals form, and many can actually form **more than one** ion!

A list of common multivalent ions is shown below:

Metal	Element Symbol	Ion Symbols	Ion names
Cobalt	Co	Co^{2+} Co^{3+}	Cobalt (II) Cobalt (III)
Chromium	Cr	Cr^{2+} Cr^{3+}	Chromium (II) Chromium (III)
Copper	Cu	Cu^{+} Cu^{2+}	Copper(I) Copper(II)
Iron	Fe	Fe^{2+} Fe^{3+}	Iron (II) Iron (III)
Lead	Pb	Pb^{2+} Pb^{4+}	Lead (II) Lead (IV)
Manganese	Mn	Mn^{2+} Mn^{4+}	Manganese (II) Manganese (IV)
Mercury	Hg	Hg^{+} Hg^{2+}	Mercury (I) Mercury (II)
Tin	Sn	Sn^{2+} Sn^{4+}	Tin (II) Tin (IV)
Titanium	Ti	Ti^{3+} Ti^{4+}	Titanium (III) Titanium (IV)
Vanadium	V	V^{3+} V^{5+}	Vanadium (III) Vanadium (V)

*Note: Some metals are found in the central block of the periodic table but **aren't multivalent**. It is helpful to remember that silver always has a charge of +1 (**Ag^{+}**) and zinc always has a charge of +2 (**Zn^{2+}**)!*

Ionic Compounds (Multivalent Metals)

Naming Ionic Compounds (Multivalent Metals)

Since these metals can form more than one possible ion, we need a naming system that shows which ion is part of the compound. Scientists have agreed to use **Roman numerals** to show the charge of the multivalent metal.

1. Name the **metal** first.
2. Then, write a **Roman numeral** in brackets to show the charge of the metal.
3. Next, use the first syllable of the non-metal + **"ide"**

Example: Ti_2O_3

*Titanium is a **transition metal**, so we know that the name will be titanium (??) oxide. In order to figure out its charge, we can use the <u>reverse</u> of our **zero sum rule** (or our **criss-cross rule**…but we need to double check that our anion charge is accurate and multiply charges as necessary!).*

↖ *#2 on p. 20 is an example of this!*

Reverse Zero-Sum Rule

O always has a charge of -2, and the subscript of 3 means the overall charge of the non-metal is -6 (i.e. 3 x (-2)).

Ti has a subscript of 2 and must have an overall charge of +6, so each Ti must have a charge of +3 (i.e. +6 ÷ 2)

∴ Ti has a charge of +3, and the name of the compound is **titanium (III) oxide**.

or

Reverse Criss-Cross Rule

$$\overset{+3}{Ti_2}\overset{-2}{O_3}$$

∴ Ti has a charge of +3, and the name of the compound is **titanium (III) oxide**.

Determining Chemical Formulae of Ionic Compounds

To determine the chemical formula, we can follow the exact same rules as we did for simple ionic compounds, except you will be given the charge of the metal in the compound name! Easy peasy!

Example: Lead (II) chloride

*see pg. 13 for a refresher on the **Zero-Sum Rule***

Practice: Naming Ionic Compounds
(Multivalent Metals)

Name each of the following compounds:

1) Fe_2O_3 _____

2) SnO_2 *see Solutions for detailed solution*   _____

3) Cr_2O_3 _____

4) $PbCl_2$ _____

5) $FeCl_2$ _____

6) CuI _____

7) MnO_2 _____

8) CuF_2 _____

9) $SnCl_4$ _____

10) CuS _____

11) PbS _____

12) TiS_2 _____

Practice: Chemical Formulae of Ionic Compounds (Multivalent Metals)

Determine the chemical formula for each of the following compounds:

1) Copper (II) chloride   _____

2) Iron (II) phosphide   _____

3) Lead (IV) oxide   _____

4) Mercury (II) oxide   _____

5) Lead (II) iodide   _____

6) Mercury (I) fluoride   _____

7) Tin (IV) sulfide   _____

8) Chromium (III) chloride   _____

9) Titanium (III) bromide   _____

10) Copper (I) oxide   _____

11) Iron (III) chloride   _____

12) Chromium (III) phosphide   _____

Self-Checking Practice: Ionic Compounds
(Multivalent Metals)

Determine the name name of each chemical compound. Highlight the correct answer and place the corresponding letter in the row of boxes at the bottom of the page from 1 to 18. A message will appear!

6	$SnCl_4$	Tin (II) chloride [J]	Tin (IV) chloride [I]
15	$HgBr_2$	Mercury (III) bromide [A]	Mercury (II) bromide [E]
5	CuI_2	Copper (II) iodide [R]	Copper (I) iodide [M]
12	Cr_2S_3	Chromium (IV) sulfide [S]	Chromium (III) sulfide [N]
9	$CuCl_2$	Copper (I) chloride [B]	Copper (II) chloride [N]
3	$PbCl_4$	Lead (II) chloride [C]	Lead (IV) chloride [P]
16	$CrCl_2$	Chromium (II) chloride [A]	Chromium (VI) chloride [R]
8	RuO_2	Ruthenium (IV) oxide [E]	Ruthenium (II) oxide [W]
18	PbI_2	Lead (II) iodide [N]	Lead (IV) iodide [D]
7	$CuCl_2$	Copper (II) chlorate [D]	Copper (II) chloride [M]
1	$CrBr_3$	Chromium (III) bromide [E]	Chromium (IV) bromine [L]
10	SnO_2	Tin (IV) oxide [T]	Tin (II) oxide [K]
17	FeS	Iron (III) sulfate [V]	Iron (II) sulfide [R]
13	Fe_2O_3	Iron (III) oxide [D]	Iron (II) oxide [E]
2	$FeCl_3$	Iron (III) chloride [X]	Iron (III) chlorate [I]
11	PbI_4	Lead (III) iodide [H]	Lead (IV) iodide [A]
14	$CrCl_3$	Chromium (III) chloride [L]	Chromium (IV) chloride [M]
4	SnO	Tin (II) oxide [E]	Tin (IV) oxide [P]

1	2	3	4	5	6	7	8	9	10	11	12	13	14	15	16	17	18

Mixed Practice
Simple Ionic Compounds + Ionic Compounds (Multivalent Metals)

Determine the corresponding chemical formula or chemical name for each of the following compounds:

1) Cr_3P_4 _____

2) $TiBr_3$ _____

3) LiI _____

4) $CaBr_2$ _____

5) $MnCl_3$ _____

6) K_3N _____

7) Sodium iodide _____

8) Zinc sulfide _____

9) Cobalt (II) chloride _____

10) Tin (II) oxide _____

11) Potassium nitride _____

12) Manganese (II) iodide _____

Ionic Compounds (Polyatomic Ions)

Sometimes ionic compounds contain **polyatomic ions.**

These are ions that are made up of many atoms that travel as a group and have a charge.

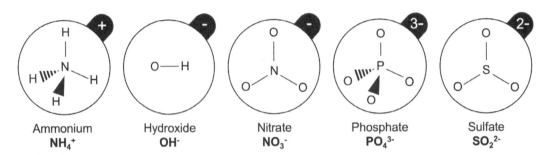

| Ammonium | Hydroxide | Nitrate | Phosphate | Sulfate |
| NH_4^+ | OH^- | NO_3^- | PO_4^{3-} | SO_2^{2-} |

Most polyatomic ions are negatively charged (i.e. they're anions), but ammonium is positively charged (i.e. it's a cation).

The table below shows some of the most common polyatomic ions:

Common Polyatomic Ions			
Name	**Ion**	**Name**	**Ion**
Acetate	$C_2H_3O^-$	Hydroxide	OH^-
Ammonium	NH_4^+	Hypochlorite	ClO^-
Bicarbonate	HCO_3^-	Nitrate	NO_3^-
Bromate	BrO_3^-	Nitrite	NO_2^-
Carbonate	CO_3^{2-}	Perchlorate	ClO_4^-
Chlorate	ClO_3^-	Permanganate	MnO_4^-
Chlorite	ClO_2^-	Phosphate	PO_4^{3-}
Chromate	CrO_4^{2-}	Phosphite	PO_3^{3-}
Cyanide	CN^-	Sulfate	SO_4^{2-}
Dichromate	$Cr_2O_7^{2-}$	Sulfite	SO_3^{2-}

Ionic Compounds (Polyatomic Ions)

Naming ionic Compounds containing Polyatomic Ions

1. Name the **cation**.
2. Then, name the **anion**.

*Usually, the anion is the polyatomic ion (*exception:* NH_4^+). Use the names of any polyatomic ions in the compound.

Example 1: **$Ba(NO_3)_2$**

barium nitrate

Example 2: **$(NH_4)_3PO_4$**

ammonium phosphate

Determining Chemical Formulae of Ionic Compounds containing Polyatomic Ions

1. Find the ionic charges using your table of polyatomic ions (pg. 24).
2. Use the **Zero-Sum Rule** (or the **Criss-Cross Rule**) to determine how many of each ion are needed to create a neutral compound.
3. Place **brackets** around the polyatomic ions with a subscript of 2 or more to show that the subscript applies to the entire polyatomic ion.

Example 1: magnesium phosphate

The phosphate ion needs brackets around it to show that the subscript applies to the whole polyatomic ion.

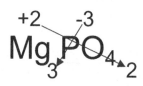

∴ $Mg_3(PO_4)_2$

Example 2: ammonium nitrate

No brackets are needed as we only have 1 of each of the polyatomic ions.

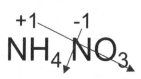

∴ NH_4NO_3

*see pg. 13 for a refresher on the **Zero-Sum Rule***

Practice: Naming Ionic Compounds
(Polyatomic Ions)

Name each of the following compounds:

1) $Mg(NO_3)_2$ _____

2) $PbSO_4$ _____

3) Na_2CO_3 _____

4) $CuSO_4$ _____

5) $LiOH$ _____

6) $Sr_3(PO_4)_2$ _____

7) KNO_3 _____

8) $Zn(CN)_2$ _____

9) $KMnO_4$ _____

10) $Ca(NO_2)_2$ _____

11) $NaHCO_3$ _____

12) $Ba(OH)_2$ _____

Practice: Chemical Formulae of Ionic Compounds (Polyatomic Ions)

Determine the chemical formula for each of the following compounds:

1) Copper (II) phosphate _____

2) Ammonium chlorate _____

3) Calcium cyanide _____

4) Aluminum sulfate _____

5) Mercury (II) sulfate _____

6) Iron (III) carbonate _____

7) Cobalt (II) nitrite _____

8) Calcium bicarbonate _____

9) Copper (I) carbonate _____

10) Iron (III) hydroxide _____

11) Sodium nitrate _____

12) Potassium chromate _____

Self-Checking Practice: Ionic Compounds
(Polyatomic Ions)

Determine the name of each compound below. Find your answer on the next page and cross out the letter above it. When you finish, the answer to the question will remain.

Na_2SO_4	$Ca_3(PO_3)_2$
$Ba(OH)_2$	NH_4Cl
$NaNO_3$	$AgNO_2$
$CaCO_3$	$NaHCO_3$
$(NH_4)_2SO_4$	$K_2Cr_2O_7$
KOH	$MgCrO_4$

Self-Checking Practice: Ionic Compounds
(Polyatomic Ions)

What do you call a cow with no legs?

L	R	G	I	E	T	R	O	S	U	D
calcium carbonate	magnesium chromate	sodium sulfite	potassium hydroxide	ammonium chloride	silver nitrite	ammonium sulfite	calcium phosphite	sodium bicarbonate	sodium nitrite	barium hydroxide

N	A	D	B	E	G	K	E	N	F	B
magnesium dichromate	calcium phosphate	silver nitrate	barium oxide	nitrogen chloride	ammonium sulfate	sodium sulfate	sodium carbonate	potassium dichromate	calcium carbide	sodium nitrate

Answer: _____

Mixed Practice
Simple Ionic Compounds, Ionic Compounds (Multivalent Metals), +
Ionic Compounds (Polyatomic Ions)

Determine the corresponding chemical formula or chemical name for each of the following compounds:

1) Rb_3P _____

2) $NaNO_3$ _____

3) Ti_2S_3 _____

4) KF _____

5) $Pb(SO_4)_2$ _____

6) Cu_3P _____

7) Lead (IV) sulfide _____

8) Zinc carbonate _____

9) Titanium (III) phosphide _____

10) Copper (II) chlorate _____

11) Sodium bromide _____

12) Vanadium (V) oxide _____

Mixed Practice (continued)

Simple Ionic Compounds, Ionic Compounds (Multivalent Metals), + Ionic Compounds (Polyatomic Ions)

Determine the corresponding chemical formula or chemical name for each of the following compounds:

13) $Fe(HCO_3)_2$ _____

14) $(NH_4)_2O$ _____

15) $Mn(NO_3)_3$ _____

16) K_2CO_3 _____

17) $CaBr_2$ _____

18) $Be(OH)_2$ _____

19) Ammonium chloride   _____

20) Copper (II) iodide   _____

21) Potassium nitride   _____

22) Copper (I) nitrite   _____

23) Aluminum fluoride   _____

24) Ammonium hydroxide   _____

Molecular Compounds

Molecules

So far, all the compounds we have explored have been ionic compounds. In ionic compounds, electrons are transferred from a metal to a non-metal to another to create an ionic bond.

When **two non-metals** bond, valence electrons are **shared** between them (*not* transferred). This type of bond is called a **covalent bond**, and the resulting compound is called a **molecular compound**.

A diagram is shown below:

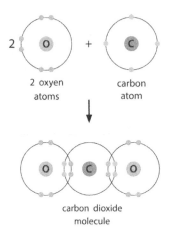

2 oxyen atoms   carbon atom

carbon dioxide molecule

When elements combine to form molecular compounds, different ratios of elements may combine to form different compounds.

For example, **carbon** and **oxygen** can form both **carbon monoxide (CO)** and **carbon dioxide (CO_2)**. These two compounds are made from the same elements but have very different properties!

Carbon dioxide	Carbon monoxide
• *a natural component of Earth's atmosphere* • *produced by various processes including cellular respiration, volcanic eruptions, and combustion of organic matter (both naturally and during industrial processes).*	• *highly toxic* • *bonds to hemoglobin in the blood, preventing oxygen from being transported to the body's cells* • *It is often produced from industrial processes and vehicle exhausts.*
$\ddot{O}\!=\!C\!=\!\ddot{O}$	$:C\!\equiv\!O:$

Molecular Compounds

Naming Molecular Compounds

Since there is no transfer of electrons in molecular compounds, **ions are not formed**.

This means that *ionic charges don't matter* with molecular compounds!

Instead, a system of prefixes is used to tell us how many of each element are present in a molecular compound.

# of atoms	Prefix
1	mono
2	di
3	tri
4	tetra
5	penta
6	hexa
7	hepta
8	octa
9	nona
10	deca

1. Name the **first non-metal** first using the corresponding **prefix**.
   Note: If there is only 1 of the first non-metal, <u>do not</u> use the prefix <u>mono.</u>
2. Name the **second non-metal** using the corresponding **prefix** and the **"ide"** ending.

Example 1: **N$_2$O$_5$**

Note: if the prefix ends in "a" and the second element starts with "o", we don't write the "a"

dinitrogen pentoxide

Example 2: **CO$_2$**

Remember, we don't write "mono" if there is 1 of the first element

carbon dioxide

Determining Chemical Formulae of Molecular Compounds

To determine the chemical formula, we can do the opposite of what we just learned in naming.

1. Write the symbol of the first element and use the **prefix** to determine the **subscript**. *Note: If there is no prefix, it means there is <u>one</u> of that element.*
2. Write the symbol of the second element and use the **prefix** to determine the **subscript**.

Example: disulfur tetrafluoride

$$S_2F_4$$

Practice: Naming Molecular Compounds

Name each of the following compounds:

1) SF_6 _____

2) Cl_2O_7 _____

3) NO _____

4) NO_2 _____

5) N_2O_3 _____

6) PCl_5 _____

7) N_2O _____

8) CO_2 _____

9) CF_4 _____

10) CCl_4 _____

11) CO _____

12) IF_7 _____

Practice: Chemical Formulae of Molecular Compounds

Determine the chemical formula for each of the following compounds:

1) Dinitrogen pentoxide   _____

2) Phosphorus pentachloride   _____

3) Nitrogen triiodide   _____

4) Tetraphosphorus decasulfide   _____

5) Chlorine monoxide   _____

6) Boron tribromide   _____

7) Disulfur tetrafluoride   _____

8) Sulfur trioxide   _____

9) Carbon monoxide   _____

10) Silicon tetrachloride   _____

11) Tetranitrogen decaselenide   _____

12) Carbon dioxide   _____

Self-Checking Practice: Molecular Compounds

Determine the chemical formula of each compound below. Find your answer on the next page and cross out the letter above it. When you finish, the answer to the question will remain.

dinitrogen monoxide	sulfur dioxide
bromine pentafluoride	diboron tetrachloride
sulfur trioxide	carbon tetrahydride
dichloride heptoxide	dihydrogen monoxide
disulfur dichloride	dinitrogen tetrahydride
dihydrogen monosulfide	carbon tetrachloride
phosphorus trichloride	

Self-Checking Practice: Molecular Compounds

What did the angry electron say when it was repelled?

L	R	F	I	E	T	O	M	S	H	D
C_4Cl	Cl_2O_7	S_2Cl_2	BrF_5	NO_2	BCl_2	SO_3	HS_2	N_2H_4	CCl_4	H_2O

E	A	P	C	T	G	K	O	N	M	B
Cl_2O_6	CH_3	PCl_3	SO_2	PCl_4	H_2S	B_2Cl_4	HO_2	N_2O	SCl	CH_4

Answer: _____

Mixed Practice

*When presented with a mixture of different types of chemical compounds, the first thing you need to do is to determine if it is **ionic** or **molecular**.*

This will dictate which naming rules you need to use.

Circle each ionic compound. Underline each molecular compound.

NO_3

$MgCl_2$

sodium chloride

magnesium sulfate

K_2SO_4

dinitrogen monoxide

Sr_3P_2

H_2O

rubidium bromide

MgF_2

$CaCO_3$

sulfur trioxide

potassium cyanide

$SiBr_4$

barium oxide

Mixed Practice

Simple Ionic Compounds, Ionic Compounds (Multivalent Metals), Ionic Compounds (Polyatomic Ions), + Molecular Compounds

Determine the corresponding chemical formula or chemical name for each of the following compounds:

1) SO_2 _____

2) K_2CO_3 _____

3) CuI _____

4) CH_4 _____

5) $Co(ClO_3)_2$ _____

6) H_2S _____

7) Nitrogen monoxide _____

8) Vanadium (V) phosphate _____

9) Tin (II) bromide _____

10) Carbon tetrabromide _____

11) Lithium cyanide _____

12) Copper (I) iodide _____

Mixed Practice (continued)
Simple Ionic Compounds, Ionic Compounds (Multivalent Metals), Ionic Compounds (Polyatomic Ions), + Molecular Compounds

13) $KClO_3$ _____

14) HgI_2 _____

15) P_2O_5 _____

16) Li_2O _____

17) $Pb(OH)_4$ _____

18) $LiBr$ _____

19) Sodium phosphide _____

20) Sulfur hexachloride _____

21) Sodium carbonate _____

22) Cesium bromide _____

23) Carbon disulfide _____

24) Iron (III) chlorite _____

Binary Acids and Oxyacids

Acids are substances that typically taste **sour** and can be **corrosive** in high concentrations. Acids play crucial roles in many important sectors and processes, including:

- *digestion in the human body;*
- *industrial processes (ex: metal etching, cleaning, etc.);*
- *the food industry;*
- *textiles (ex; used in dyeing and processing); and*
- *lots more!*

Examples of acids include hydrochloric acid (found in gastric juice), sulfuric acid (used in car batteries), and citric acid (found in citrus fruits).

Acids are compounds that *dissociate in water to produce hydrogen ions* [H+].

This means that if a chemical formula **starts with H** and the compound is in an **aqueous state (aq)**, we need to use the rules for *naming acids*.

The two main categories of acids are **binary acids** and **oxyacids**.

Binary Acids

A **binary acid** is an acid that consists of **hydrogen** and **one other element** (often a halogen).

Naming Binary Acids

1. First, write the prefix **hydro-**
2. Then, write the base name of the anion with the suffix **–ic** and the word **acid.**

Example 1: $HBr_{(aq)}$

Hydrobromic acid

Example 2: $HI_{(aq)}$

Hydroiodic acid

Binary Acids and Oxyacids

Oxyacids

An **oxyacid** is an acid that is made up of **hydrogen**, **oxygen**, and a **non-metal.**

Oxyacids are often formed with a **polyatomic ion** that reacts with **hydrogen**. The polyatomic ion is also called the **oxyanion** when it is part of an oxyacid.

Naming Oxyacids

1. If the oxyanion ends with **–ite**, add the suffix **–ous** to its base name of the word **acid.**
2. If the oxyanion ends with **–ate**, add the suffix **–ic** to its base name of the word **acid.**

Example 1: $HNO_{3(aq)}$

Nitric acid

Example 2: $H_2SO_{3(aq)}$

sulfurous acid

Determining Chemical Formulae of Binary Acids and Oxyacids

To determine the chemical formula of both **binary acids** and **oxyacids**, we can follow the same general rules as we did for ionic compounds.

The first element will always be hydrogen, which will have a charge of **+1**. The <u>prefix and suffix </u>of the acid name will guide you to the anion in the compound.

> **hydro- + -ic** = *one non-metal*
> **-ic** = *polyatomic ion ending in "ate"*
> **-ous** = *polyatomic ion ending in "ite"*

Example 1: <u>hydro</u>sulfur<u>ic</u> acid

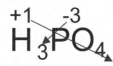

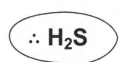

Example 2: phosphor<u>ic</u> acid

*see pg. 13 for a refresher on the **Zero-Sum Rule***

Practice: Naming Binary Acids and Oxyacids

*Name each of the following acids. *Assume they are all in an aqueous state.*

1) H_3PO_4 _____

2) HI _____

3) $HClO_3$ _____

4) HF _____

5) $H_2Cr_2O_4$ _____

6) HNO_3 _____

7) HBr _____

8) H_2CO_3 _____

9) H_2Se _____

10) H_2SO_4 _____

11) H_2S _____

12) H_3P _____

Practice: Chemical Formulae of Binary Acids and Oxyacids

Determine the chemical formula for each of the following acids:

1) Hydrochloric acid   _____

2) Sulfuric acid   _____

3) Hydrosulfuric acid   _____

4) Hydroiodic acid   _____

5) Chlorous acid   _____

6) Carbonic acid   _____

7) Acetic acid   _____

8) Nitric acid   _____

9) Hydrobromic acid   _____

10) Phosphoric acid   _____

11) Nitrous acid   _____

12) Phosphorous acid   _____

Self-Checking Practice: Binary Acids and Oxyacids

Determine the name name of each acid. Highlight the correct answer and place the corresponding letter in the row of boxes at the bottom of the page from 1 to 16. A message will appear! *All compounds below are assumed to be aqueous.

4	H_2SO_4	Sulfurous acid **[N]**	Sulfuric acid **[A]**
5	HCl	Hydrochloric acid **[M]**	Hydrochlorous acid **[P]**
13	H_2CO_3	Carbonic acid **[O]**	Hydrocarbic acid **[A]**
1	HF	Hydrofluoric acid **[D]**	Fluorous acid **[K]**
9	H_2S	Hydrosulfuric acid **[A]**	Sulfurous acid **[H]**
6	H_2SO_3	Sulfuric acid **[H]**	Sulfurous acid **[B]**
14	H_3PO_4	Phosphoric acid **[W]**	Hydrophosphoric acid **[U]**
2	$HClO_2$	Hydrochloric acid **[V]**	Chlorous acid **[R]**
11	HBr	Hydrobromic acid **[T]**	Bromic acid **[O]**
7	HNO_2	Nitrous acid **[I]**	Nitric acid **[W]**
12	HI	Hydroiodic acid **[N]**	Iodous acid **[C]**
10	$HClO_4$	Perchloric acid **[C]**	Hydrochloric acid **[D]**
8	HNO_3	Nitrous acid **[N]**	Nitric acid **[G]**
3	$HBrO_3$	Hydrobromic acid **[P]**	Bromic acid **[E]**

1	2	3	4	5	6	7	8	9	10	11	12	13	14

Mixed Practice 1

Simple Ionic Compounds, Ionic Compounds (Multivalent Metals), Ionic Compounds (Polyatomic Ions), Molecular Compounds, + Acids

Name each of the following compounds:

1) $CuSO_4$ _____

2) $H_2SO_{4(aq)}$ _____

3) $RbCN$ _____

4) SiF_4 _____

5) $CsBr$ _____

6) $LiOH$ _____

7) SO_3 _____

8) $Cu(OH)_2$ _____

9) $HCl_{(aq)}$ _____

10) NH_4OH _____

11) Ba_3N_2 _____

12) $Sn(NO_3)_4$ _____

Mixed Practice 2

Simple Ionic Compounds, Ionic Compounds (Multivalent Metals), Ionic Compounds (Polyatomic Ions), Molecular Compounds, + Acids

Name each of the following compounds:

13) CCl_4 _____

14) $Sr_3(PO_4)_2$ _____

15) BaO _____

16) $Fe(OH)_2$ _____

17) BeS _____

18) $H_2CO_{3(aq)}$ _____

19) S_2Cl_2 _____

20) Na_2SO_4 _____

21) CuF _____

22) $HBr_{(aq)}$ _____

23) $Ba(OH)_2$ _____

24) NO _____

Mixed Practice 3

Simple Ionic Compounds, Ionic Compounds (Multivalent Metals), Ionic Compounds (Polyatomic Ions), Molecular Compounds, + Acids

Determine the chemical formula for each of the following compounds:

1) Tin (II) sulfate _____

2) Carbon monoxide _____

3) Strontium chlorate _____

4) Nitrogen dioxide _____

5) Magnesium phosphide _____

6) Phosphorus triiodide _____

7) Hydroiodic acid _____

8) Aluminum phosphide _____

9) Copper (I) sulfite _____

10) Arsenic tribromide _____

11) Phosphoric acid _____

12) Calcium nitride _____

Mixed Practice 4

Simple Ionic Compounds, Ionic Compounds (Multivalent Metals), Ionic Compounds (Polyatomic Ions), Molecular Compounds, + Acids

Determine the chemical formula for each of the following compounds:

13) Ammonium sulfate _____

14) Nitric acid _____

15) Zinc phosphate _____

16) Mercury (I) cyanide _____

17) Silicon dioxide _____

18) Aluminum chlorate _____

19) Chlorous acid _____

20) Chlorine dioxide _____

21) Ammonium phosphate _____

22) Lead (IV) dichromate _____

23) Diphosphorus trioxide _____

24) Hydroiodic acid _____

SOLUTIONS

Solutions

Practice: Who am I?
(p. 8)
1) Magnesium (Mg)
2) Hydrogen (H)
3) Helium (He)
4) Chlorine (Cl)
5) Hafnium (Hf)
6) Chlorine (Cl)
7) Uranium (U)
8) Krypton (Kr)
9) Beryllium (Be)
10) Sodium (Na)
11) Silicon (Si)
12) Calcium (Ca)

Practice: Chemical Symbols
(p. 10)
1) Sodium – Na
2) Tin – Sn
3) Copper – Cu
4) Mercury – Hg
5) Potassium – K
6) Lead – Pb
7) Antimony – Sb
8) Silver – Ag
9) Gold – Au
10) Strontium – Sr
11) Phosphorus – P
12) Nitrogen – N

Practice: Metal or Non-metals
(p. 11)
Metals – lithium, Be, magnesium, Cu, Rb, Ti, tungsten, Pb
Non-metals/metalloids – Cl, F, phosphorus, oxygen, H, S, nitrogen, carbon, oxygen, sulfur, Si, helium

Practice: Naming Simple Ionic Compounds
(p. 14)
1) Sodium chloride
2) Calcium fluoride
3) Lithium bromide
4) Aluminum sulfide
5) Magnesium nitride
6) Potassium oxide
7) Strontium phosphide
8) Barium iodide
9) Sodium bromide
10) Rubidium oxide
11) Lithium sulfide
12) Magnesium bromide

Practice: Chemical Formulae of Simple Ionic Compounds
(p. 10)
1) MgS
2) Na_3P
3) Li_2O
4) AlN
5) KBr
6) $SrCl_2$
7) Rb_3P
8) NaI
9) BeO
10) AlF_3
11) MgO
12) Ca_3P_2

Self-Checking Practice:
Simple Ionic Compounds
(p. 16/17)

Answer: a palm tree

Practice: Naming Ionic Compounds (Multivalent Metals)
(p. 20)
1) Iron (III) oxide
2) Tin (IV) oxide*
3) Chromium (III) oxide
4) Lead (II) chloride
5) Iron (II) chloride
6) Copper (I) iodide
7) Manganese (IV) oxide
8) Copper (II) fluoride
9) Tin (IV) chloride
10) Copper (II) sulfide
11) Lead (II) sulfide
12) Titanium (IV) sulfide

Full solution: Finding the name of 2) SnO_2 using the Reverse Criss-Cross Rule

If we do a reverse criss-cross for this compound, it appears as if the charge on the oxide ion is -1.

We know from the position of oxygen on the periodic table that it will always have a charge of -2 (NOT -1)

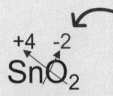

This means that we need to multiply both charges by 2. These charges now match what we know about oxygen and also satisfy the Zero-Sum rule.

∴the name of the compound is **tin (IV) oxide**.

Solutions (continued)

Practice: Chemical Formulae of Ionic Compounds (Multivalent Metals) (p. 21)
1) $CuCl_2$
2) Fe_3P_2
3) PbO_2
4) HgO
5) PbI_2
6) HgF
7) SnS_2
8) $CrCl_3$
9) $TiBr_3$
10) Cu_2O
11) $FeCl_3$
12) CrP

Self-Checking Practice: Ionic Compounds (Multivalent Metals) (p. 22)

Answer: Experiment and learn

Mixed Practice (p. 23)
1) Chromium (IV) phosphide
2) Titanium (III) bromide
3) Lithium iodide
4) Calcium bromide
5) Manganese (III) chloride
6) Potassium nitride
7) NaI
8) ZnS
9) $CoCl_2$
10) SnO
11) K_3N
12) MnI_2

Practice: Naming Ionic Compounds (Polyatomic Ions) (p. 26)
1) Magnesium nitrate
2) Lead (II) sulfate
3) Sodium carbonate
4) Copper (II) sulfate
5) Lithium hydroxide
6) Strontium phosphate
7) Potassium nitrate
8) Zinc cyanide
9) Potassium permanganate
10) Calcium nitrite
11) Sodium bicarbonate
12) Barium hydroxide

Practice: Chemical Formulae of Ionic Compounds (Polyatomic Ions) (p. 27)
1) $Cu_3(PO_4)_2$
2) NH_4ClO_3
3) $Ca(CN)_2$
4) $Al_2(SO_4)_3$
5) HgS
6) $Fe_2(CO_3)_3$
7) $Co(NO_2)_2$
8) $Ca(HCO_3)_2$
9) Cu_2CO_3
10) $Fe(OH)_3$
11) $NaNO_3$
12) K_2CrO_4

Self-Checking Practice: Ionic Compounds (Polyatomic Ions) (p. 28/29)

Answer: ground beef

Mixed Practice (p. 30)
1) Rubidium phosphide
2) Sodium nitrate
3) Titanium (III) sulfide
4) Potassium fluoride
5) Lead (IV) sulfate
6) Copper (I) phosphide
7) PbS_2
8) $ZnCO_3$
9) TiP
10) $CuClO_3$
11) $NaBr$
12) V_2O_5

Mixed Practice (p. 31)
13) Iron (II) bicarbonate
14) Ammonium oxide
15) Manganese (III) nitrate
16) Potassium carbonate
17) Calcium bromide
18) Beryllium hydroxide
19) NH_4Cl
20) CuI_2
21) K_3N
22) $CuNO_2$
23) AlF_3
24) NH_4OH

Practice: Naming Molecular Compounds (p. 34)
1) Sulfur hexafluoride
2) Dichlorine heptoxide
3) Nitrogen monoxide
4) Nitrogen dioxide
5) Dinitrogen trioxide
6) Phosphorus pentachloride
7) Dinitrogen monoxide
8) Carbon dioxide
9) Carbon tetrafluoride
10) Carbon tetrachloride
11) Carbon monoxide
12) Iodine heptafluoride

Solutions (continued)

Practice: Chemical Formulae of Molecular Compounds (p. 35)
1) N_2O_5
2) PCl_5
3) NI_3
4) P_4S_{10}
5) ClO
6) BBr_3
7) S_2F_4
8) SO_3
9) CO
10) $SiCl_4$
11) N_4Se_{10}
12) CO_2

Mixed Practice (p. 39)
1) Sulfur dioxide
2) Potassium carbonate
3) Copper (I) iodide
4) Carbon tetrahydride
5) Cobalt (II) chlorate
6) Dihydrogen monosulfide
7) NO
8) $V_3(PO_4)_5$
9) $SnBr_2$
10) CBr_4
11) $LiCN$
12) CuI

Practice: Chemical Formulae of Acids (p. 45)
1) HCl
2) H_2SO_4
3) H_2S
4) HI
5) $HClO_2$
6) H_2CO_3
7) HC_2H_3O
8) HNO_3
9) HBr
10) H_3PO_4
11) HNO_2
12) H_3PO_3

Self-Checking Practice: Molecular Compounds (p. 36/37)

Answer: Let me atom

Mixed Practice (p. 40)
13) Potassium chlorate
14) Mercury (II) iodide
15) Diphosphorus pentoxide
16) Lithium oxide
17) Lead (IV) hydroxide
18) Lithium bromide
19) Na_3P
20) SCl_6
21) Na_2CO_3
22) $CsBr$
23) CS_2
24) $Fe(ClO_2)_3$

Self-Checking Practice: Acids (p. 46)

Answer: Dream big, act now

Mixed Practice (p. 38)

Ionic compounds: sodium chloride, $MgCl_2$, K_2SO_4, magnesium sulfate, Sr_3P_2, rubidium bromide, MgF_2, $CaCO_3$, barium oxide, potassium cyanide

Molecular compounds: NO_3, dinitrogen monoxide, H_2O, sulfur trioxide, $SiBr_4$

Practice: Naming Acids (p. 44)
1) Phosphoric acid
2) Hydroiodic acid
3) Chloric acid
4) Hydrofluoric acid
5) Chromic acid
6) Nitric acid
7) Hydrobromic acid
8) Carbonic acid
9) Hydroselenic acid
10) Sulfuric acid
11) Hydrosulfuric acid
12) Hydrophosphoric acid

Mixed Practice 1 (p. 47)
1) Copper (II) sulfate
2) Sulfuric acid
3) Rubidium cyanide
4) Silicon tetrafluoride
5) Cesium bromide
6) Lithium hydroxide
7) Sulfur trioxide
8) Copper (II) hydroxide
9) Hydrochloric acid
10) Ammonium hydroxide
11) Barium nitride
12) Tin (IV) nitrate

Solutions (continued)

Mixed Practice 2
(p. 48)

13) Carbon tetrachloride
14) Strontium phosphate
15) Barium oxide
16) Iron (II) hydroxide
17) Beryllium sulfide
18) Carbonic acid
19) Disulfur dichloride
20) Sodium sulfate
21) Copper (I) fluoride
22) Hydrobromic acid
23) Barium hydroxide
24) Nitrogen monoxide

Mixed Practice 3
(p. 49)

1) $SnSO_4$
2) CO
3) $Sr(ClO_3)_2$
4) NO_2
5) Mg_3P_2
6) PI_3
7) HI
8) AlP
9) Cu_2SO_3
10) $AsBr_3$
11) H_3PO_4
12) Ca_3N_2

Mixed Practice 4
(p. 50)

13) $(NO_4)_2SO_4$
14) HNO_3
15) $Zn_3(PO_4)_2$
16) $HgCN$
17) SiO_2
18) $Al(ClO_3)_3$
19) $HClO_2$
20) ClO_2
21) $(NH_4)_3PO_4$
22) $Pb(Cr_2O_7)_2$
23) P_2O_3
24) HI

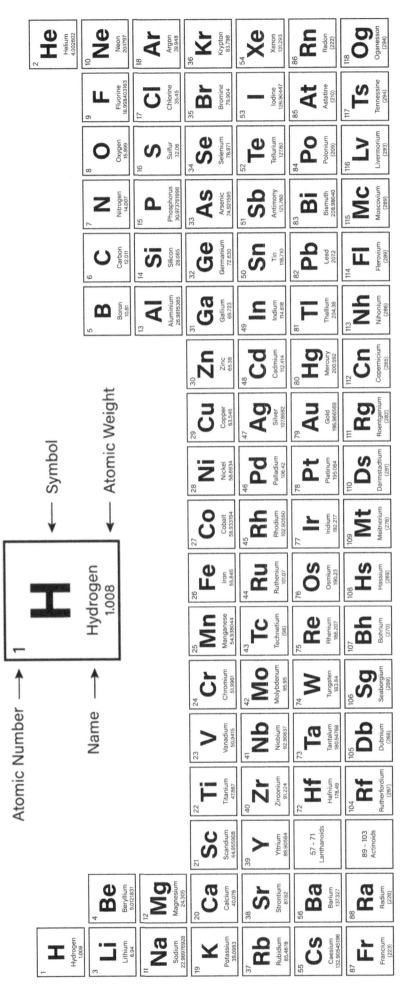

Quick Reference Sheet

Common Multivalent Metals

Metal	Element Symbol	Ion Symbols	Ion names
Cobalt	Co	Co^{2+} Co^{3+}	Cobalt (II) Cobalt (III)
Chromium	Cr	Cr^{2+} Cr^{3+}	Chromium (II) Chromium (III)
Copper	Cu	Cu^{+} Cu^{2+}	Copper(I) Copper(II)
Iron	Fe	Fe^{2+} Fe^{3+}	Iron (II) Iron (III)
Lead	Pb	Pb^{2+} Pb^{4+}	Lead (II) Lead (IV)
Manganese	Mn	Mn^{2+} Mn^{4+}	Manganese (II) Manganese (IV)
Mercury	Hg	Hg^{+} Hg^{2+}	Mercury (I) Mercury (II)
Tin	Sn	Sn^{2+} Sn^{4+}	Tin (II) Tin (IV)
Titanium	Ti	Ti^{3+} Ti^{4+}	Titanium (III) Titanium (IV)
Vanadium	V	V^{3+} V^{5+}	Vanadium (III) Vanadium (V)

Common Polyatomic Ions

Name	Ion	Name	Ion
Acetate	$C_2H_3O^-$	Hydroxide	OH^-
Ammonium	NH_4^+	Hypochlorite	ClO^-
Bicarbonate	HCO_3^-	Nitrate	NO_3^-
Bromate	BrO_3^-	Nitrite	NO_2^-
Carbonate	CO_3^{2-}	Perchlorate	ClO_4^-
Chlorate	ClO_3^-	Permanganate	MnO_4^-
Chlorite	ClO_2^-	Phosphate	PO_4^{3-}
Chromate	CrO_4^{2-}	Phosphite	PO_3^{3-}
Cyanide	CN^-	Sulfate	SO_4^{2-}
Dichromate	$Cr_2O_7^{2-}$	Sulfite	SO_3^{2-}

Want a **FREE** resource to help you master chemical reactions?

Made in the USA
Las Vegas, NV
03 May 2024

89476612R00033